Kamathe Katsongo

Essay on the epistemology of planning disciplines

Kamathe Katsongo

Essay on the epistemology of planning disciplines

ScienciaScripts

Cover image: www.ingimage.com

This book is a translation from the original published under ISBN 978-620-6-72033-1.

Publisher:
Sciencia Scripts
is a trademark of
Dodo Books Indian Ocean Ltd. and OmniScriptum S.R.L publishing group

120 High Road, East Finchley, London, N2 9ED, United Kingdom
Str. Armeneasca 28/1, office 1, Chisinau MD-2012, Republic of Moldova, Europe
Printed at: see last page
ISBN: 978-620-8-07920-8

An essay on the epistemology of planning disciplines*.

By: Kamathe KATSONGO, PhD

Table of contents

Introduction

Where does planning begin and end as a field of research and action? Answering this question is a challenge, probably even more so in the first case (field of research) than in the second (field of action), so much so that planning is sometimes confused with urbanism, architecture, geography, civil engineering, design (interior and industrial), planning and so on.

This difficulty is compounded by the difficulty of defining, or rather identifying precisely, the theoretical frameworks that underpin planning research, which seems to be more a matter of picking and choosing than of drawing on a clearly identified corpus. It is in this context that we draw on three theses from different disciplines to better define the specificity of planning disciplines.

In order to gain a better understanding of the nature and content of theses from different disciplines, which are essential for grasping the characteristics of planning disciplines, and thus to be able to position ourselves in an open and rigorous epistemological debate on spatial planning, we will first present the main epistemological trends, and then focus our analysis on the content of theses from the Faculty of Law, the Faculty of Pharmacy and the Faculty of Planning.

For each thesis, we'll begin with a brief presentation of its title, author, faculty of origin and content. We then turn to the problematic. At this level, the emphasis will be on the epistemological posture adopted by the author, the construction of the argument in relation to the theoretical and methodological frameworks, and the author's positioning in relation to the advancement of knowledge, with reference to established and accepted models.

Finally, we will address the main research question, emphasizing its representativeness in relation to the epistemological stance adopted, in an attempt to identify how it fits into the general problematic.

1. Concept of epistemology

The word **epistemology** comes from the Greek *episteme̊* ("knowledge", "science") and *logos* ("discourse").**epistemology** is a reflection on science: it is a part of philosophy concerned with knowledge as such, a rational discourse on scientific knowledge. Philosophy of science thus studies scientific knowledge from a critical point of view. What is knowledge? How is it acquired? How do we use it? Why seek it or use it? Epistemology may be concerned with establishing a classification of the sciences, defining categories.

In everyday language, the word *science* can have several meanings, and before embarking on a discussion of epistemology (from the Greek *epistêmê* "science" and *logos* "study"), it's a good idea to differentiate between them. According to Robert (1995, p. 2051), quoted by Riopel (2005), in its broadest application, the word *science* is often confused with the word *savoir* or even simply *connaissance*. This definition, which is too broad, is certainly not the one we wish to adopt for the purposes of this paper. Again according to the same author, the word *science* can also be associated with the *know-how that comes from knowledge*, and although this meaning is already more restrictive, it still doesn't suit us.

Instead, we will use the following definition proposed by Robert, specifying that it is the modern and common meaning:

> "*Body of knowledge, studies of universal value, characterized by a specific object (field) and method, and based on verifiable objective relationships.*" *(p. 2051)*

Granger (1995, pp. 45-48) takes up the essence of the previous definition, separating it into three main features characterizing science, which we summarize as follows:

> *1. science aims at a reality through a constant, laborious and yet creative search for concepts oriented towards the description or organization of data that resist our whims;*
> *2. the ultimate aim of science is to describe, explain and understand, not directly to act;*
> *3. science is constantly concerned with producing criteria for public validation, i.e. exposed to anyone's informed scrutiny.*

With his criterion of *refutability,* Popper (1985, p.230) goes even further, proposing that a body of knowledge, to qualify as science, must not only be verified or verifiable, but must also be exposed in advance to being refuted by experience (by experience, we mean the result of interaction with reality). Popper's particularly restrictive criterion involves two main ideas: firstly, the need for a scientific theory to make at least one prediction, and secondly, the need for this prediction to relate to a new experiment (the outcome of which is not yet known with sufficient precision) likely to disprove the theory. By adding a temporal dimension to the definition of a science, Popper excludes from the realm of science, among other things, all theories that merely adjust *a posteriori* to experiments, predicting nothing new.

Obviously, this definition of science does not meet with unanimous approval. Jarroson (1992, pp. 167-168) presents three limits to the use of Popper's criterion, which we summarize as follows:

1. There are propositions that make sense, but cannot be disproved. For example, "there are immortal men"; you would have to kill all men to prove that this proposition is false.
2. It is rare for an experiment to refute only one theory at a time. For example, when we observe a falling marble to study mechanics, we also admit the theory of light that allows us to see the marble.
3. You can never be sure of the validity of an experiment or a set of experiments. We must always make the fundamental conjecture of relying on experience.

The first limit poses no major problem, since we can always restrict ourselves to considering only those theories that are effectively refutable. The second limit forces us to consider science as a whole, which is not an insurmountable problem. However, the third and more profound limitation deserves our attention.

The impossibility of establishing objective relationships obviously invalidates the objective process of verification, and makes it impossible to apply Popper's criterion strictly. The definition of science must therefore be revised.

2. Overview of the main epistemological trends

In the following sections, we present the main epistemological currents concerning scientific activity, as listed in Bégin (1997, p. 13) or Alters (1997, p. 50-53): rationalism, empiricism, positivism, constructivism and realism. For each of these currents, we will attempt to give a sufficiently precise definition and an outline of the implications of this definition for the search for new knowledge, along with the names of a few scientists who have defended positions consistent with the definition adopted.

2.1. Rationalism (17th century)

According to *Le dictionnaire actuel de l'éducation* (1994, p. 1003), rationalism is an epistemological trend that considers that "*all valid knowledge comes either exclusively or essentially from the use of reason*". According to Bégin (1997, p. 12), Alters (1997, p. 51) and Blanshard (2001), Greek philosophers such as Euclid (~300 B.C.), Pythagoras (569-475 B.C.) and Plato (428-347 B.C.) are generally acknowledged to have defended rationalist positions, giving primacy to ideas. This association is obviously *a posteriori*, since the rationalist epistemological trend had not yet been defined at the time. More recently, the mathematicians Descartes (1596-1650) and Leibniz (1646-1716), as well as the philosopher Kant (1724-1804), have been associated with this trend, which favours reasoning in general, and deductive (or analytical) reasoning in particular, which moves from the abstract to the concrete as a mechanism for producing knowledge.

Historically, the knowledge associated with the field of geometry has played an important role in the development and justification of the rationalist epistemological position. For example, Britannica (2001) reports that Plato, in his dialogue entitled *Meno*, highlights the certain, universal and innate nature of knowledge by recounting how Socrates managed to get an illiterate young slave to demonstrate, step by step and without teaching him, the Pythagorean theorem applied to the diagonal of a square. Later, in the early 17th century, the inventor of analytic geometry, French mathematician René Descartes, took up the rationalist position, attempting to apply the rigor and clarity of mathematics to the realm of philosophy. In the same spirit, the Italian physicist and astronomer Galileo (1564-1642), while recognizing the importance of experimentation and observation in the search for new knowledge (having himself made seminal astronomical observations of the moons of Jupiter and the phases of Venus), proposed in *The Essayist* in 1623 that :

> *"The great book of the Universe is written in the language of mathematics. This book can only be understood if we first learn its language, and the alphabet in which it is written. The characters are triangles and circles, as well as the other geometrical figures without which it is humanly impossible to decipher a single word"*.

Within the rationalist trend, we can distinguish, among others, *Platonism*, which believes, according to Barreau (1995, p. 50), "*in an inherent harmony in nature that reflects itself in our minds*", from the *criticism* of Kant (1724-1804), who considers that knowledge depends on structures inscribed *a priori* in the human mind that make it possible to perceive reality.

2.2 Empiricism (18^e^ century)

Empiricism is radically opposed to rationalism, proposing that *all knowledge comes essentially from experience.* Popper (1985, p. 217) recognizes *a posteriori* an empiricist way of thinking in the generalizing propositions of the Greek philosopher Anaximenes (610-545 BC). More recently, according to Quinton (2001), the English philosophers Bacon (1561-1626), Locke (1632-1704) and Berkeley (1685-1753) are associated with this trend, which proposes that the sciences progress by accumulating observations from which laws can be extracted through inductive (or synthetic) reasoning that moves from the concrete to the abstract. For empiricists, observations are the key to understanding reality.

As Riopel (2005) points out, for empiricists, deduction is excluded from the mechanism for producing new knowledge. Deduction is merely a temporary step in making a hypothesis or simplifying the description of all the observations made by scientists at a given time. Empiricists are more flexible in their definition of *reasoning*, especially when it comes to inductive reasoning.

Historically, the work of Newton (1642-1726), with its emphasis on experimentation, contributed significantly to the spread of the empiricist position. In the preface to his *Principia* in 1686, Newton proposed that observation of phenomena should usually precede demonstration:

> *"the whole burden of philosophy seems to consist in this - from the phenomena of motions to investigate the forces of nature, and then from these forces to demonstrate the other phenomena".*

The application of this method enabled Newton and his contemporaries to describe the forces in mechanics (particularly gravity) and to build a corpuscular model of light. A little later, Coulomb (1736-1806) used the same method to demonstrate the electrical force. In chemistry, Lavoisier (1743-1794) drew on Newton's work on light to lay the foundations of modern chemistry by proposing an experimental method for identifying the fundamental elements.

2.3. Positivism (19^e^ century)

Although Feigl (2001) mentions that the Greek philosopher Sextus Empiricus (160-210), who lived at the turn of the 3^e^ century, adopted a positivist stance by insisting on the suspension of all judgment, the positivist current is generally attributed to the philosopher Auguste Comte (1718-1857) as well as to the physicists Mach (1838-1916), Bridgman (1882-1961) and Bohr (1885-1962). The positivist movement is inspired by empiricism in that it focuses on observational facts alone, but recognizes the importance of reasoning, adding that the sciences use mathematization to link experimental data together as simply as possible (Bégin, 1997, p. 12).

This marriage between reasoning and experience was already clear in 1820 in Auguste Comte's definition, quoted by Kremer-Marietti (1993, p. 6):

> "*If it is true that a science becomes positive only when it is based exclusively on observed facts whose accuracy is generally recognized, it is equally indisputable [...] that any branch of our knowledge becomes a science only when, by means of a hypothesis, we have linked all the facts that serve as its basis.*"

It should be noted that positivists insist on the rigor of inductive reasoning, which enables us to move from facts to hypotheses. Positivists such as the philosopher and economist Stuart Mill (1806-1873) and the geneticist Fisher (1890-1962) developed inductive methods, based on probability and statistics, to obtain probable laws from a set of measurements. However, according to Barreau (1995, p. 56), to this day, there is no strict inductive logic that does not contain a purely conventional component. As inductive reasoning is indispensable (for positivists) to the evolution of science (according to Auguste Comte's famous formula "*voir pour prévoir*"), the theories produced have no value in themselves other than that of being linked to facts. They tell us nothing about reality that is not already contained in the facts themselves. Consequently, for positivists, "*science describes the how of things without being able to say anything about their why*" (*Le dictionnaire actuel de l'éducation*, 1994, p. 1003).

Historically, this clear distinction between observations (*the how*) and mathematical models (*the why*) is particularly important in understanding what led positivists to distinguish themselves from empiricists. For example, the experimental work of Dalton (1766-1844), which founded chemical atomism, raised the fundamental question of whether atoms really existed. The empiricists of the time generally believed that atoms, since they were necessary to explain experimental results, really did exist. Positivists fiercely opposed the existence of atoms because they were not directly observable: atoms were models (*the why*) to explain experiments (*the how*). For positivists, models are human creations that have strictly no value other than being useful. Positivists thus categorically opposed anything in scientific models that was not directly observable.

According to Kremer-Marietti (1993, pp. 10-11), within the positivist trend, we can distinguish between the *conventionalism* of Poincarré (1854-1912), who proposes that *hypotheses have no cognitive value in themselves*, the *pragmatism* of James (1842-1910), who proposes, according to Le Moigne (1995, p.. 55), that "*the true consists simply in what is advantageous for thought*", and the *logical positivism* of Carnap (1891-1970), which proposes that *the cognitive processes involved in the elaboration of representations must be capable of being constructed or reconstructed.* Logical positivism is sometimes presented as one of the precursors of constructivism.

2.4. Constructivism (20th^e century)

According to Le Moigne (1995, p. 43), certain ideas can be found among the Greek sophists that can be associated *a posteriori* with the heritage of the constructivist position. He cites, for example, Heraclitus' (550-480 BC) conception of the ambiguity of reality, and Protagoras' (485-410 BC) formula: "*Man is the measure of all things*" (p. 43). It was in the 20th century, however, that the constructivist movement emerged, and is usually credited to the Dutch mathematician Brouwer (1881-1966), who used the term *constructivist* to characterize his position on the question of foundations in mathematics, in opposition to Hilbert's *formalist* position (Largeault, 1992, p. 27).

The constructivist position was taken up by the Swiss psychologist Piaget (1896-1980), in the 1960s, to explain the basis of knowledge. Indeed, Piaget and Garcia (1983, p. 30) propose that "*a fact is [...] always the product of composition, between a part supplied by objects, and another constructed by the subject*". By insisting on the highly constructed nature of knowledge in general, and scientific knowledge in particular, the constructivist current calls into question the possibility of always obtaining objective relationships on which to base science. The absence of objective relationships obviously invalidates any formal verification process, and makes the strict application of Popper's criterion (as presented above) impossible. By renouncing objectivity, the constructivist current considers scientific knowledge in the same way as any other knowledge, and proposes that the sciences *construct* (rather than reveal) a possible reality on the basis of successive cognitive experiences.

According to Larochelle and Désautels (1992, p. 27), constructivists do not reject the existence of an ultimate reality, but they do assert that it cannot be known. To illustrate this point, the authors use the analogy of the key:

> *"Knowledge fits reality like a key fits a lock. Suitability applies to the key, not to the lock. In other words, I can describe the key without being able to describe the lock. [...] Just as the key does not reproduce the lock, neither does knowledge reproduce reality." (p. 27-28)*

2.5. Realism (20th^{e} century)

According to Bégin (1997, p. 13), the Greek philosopher Aristotle (384-322 B.C.), in his concern to build some of his models from systematic observations of nature, defended a position that can be described *a posteriori* as realist. Realism proposes that scientific models are approximations of an objective reality that exists independently of the observer. In contrast to rationalism, empiricism and positivism, this current does not retain a precise mechanism for the progression of knowledge, but rather recognizes the complementarity of different approaches. According to Owens and Starkey (2001), the physicists Planck (1858-1947) and Einstein (1879-1955) are generally associated with this current, considering the latter's reaction: "*God doesn't play dice!*" to the positivist position of Bohr (1885-1962), who claimed that quantum uncertainty, since it is always measured, is an intrinsic property of reality.

It is the recognition of the existence of a reality towards which scientific models (which are, moreover, human constructs) tend that distinguishes realism from constructivism. In contrast to the radical constructivist proposition that the observer *constructs reality*, realism proposes that *the observer is part of reality*. The distinction is important, since the realist position, while recognizing the highly constructed nature of scientific knowledge, admits a mechanism of selection of this knowledge that corresponds to interaction with reality with the aim of predicting it. In this way, Popper's criterion can be rigorously applied, even though constructed science itself has a subjective component. In other words, reality reacts coherently (insofar as reality is coherent) regardless of the model chosen to describe it.

It is important to note that the realist position, by not adopting a precise mechanism for the construction of scientific knowledge, cannot claim that science is progressing in its attempt to describe reality. At best, the accuracy of predictions increases, but the construction of the model that produces these predictions remains just as subjective as before. Thus, scientific theories are inscribed within broad *paradigms* that result from the implicit and subjective agreement of the scientific community (Kuhn, 1983, explained by Jarroson, 1992, p. 171).

Historically, the work of Michelson (1852-1931) on the speed of light and that of Einstein on relativity in 1905 helped to diminish the influence of the positivist position (in favor of the realist position) by seriously calling into question the necessity of the notion of ether hitherto defended by the positivists. Similarly, the work of Rutherford (1871-1937) on the atomic nucleus and that of Bohr (1885-1962) on electron orbits around the nucleus reinforced the hypothesis of the real existence of atoms, which positivists had opposed from the outset. In this context, the realist position distinguished itself from the positivist position by recognizing a certain reality in the models developed, which were intended to be increasingly accurate approximations of a single reality.

Table I. Epistemological trends

Current description	Pedagogical trends	Philosopher or scientist
Rationalism (17e century) All valid knowledge comes essentially from the use of reason.	Emphasize rationalization over experimentation.	Plato (428-347 BC) Descartes (1596-1650) Leibnitz (1646-1716) Kant (1724-1804)
Empiricism (18e century) All valid knowledge comes essentially from experience.	Emphasize the importance of experimentation over rationalization.	Anaximenes (610-545 BC) Bacon (1561-1626) Locke (1632-1704) Newton (1642-1726) Berkeley (1685-1753)

Current description	Pedagogical trends	Philosopher or scientist
Positivism (19e century) Science progresses on the basis of measured facts, from which it extracts models through rigorous inductive reasoning. Anything that cannot be directly measured does not exist.	Recognize the complementary importance of experimentation and rationalization, emphasizing the scientific approach that drives science forward.	Sextus Empiricus (160-210) Comte (1718-1857) Stuart Mill (1806-1873) Mach (1838-1916) Bridgman (1882-1961) Bohr (1885-1962) Carnap (1891-1970)
Constructivism (20the century) Scientific knowledge (observations and models) are subjective constructs that tell us nothing about reality.	Emphasize the arbitrary or subjective nature of scientific models, encouraging students to construct their own knowledge.	Heraclitus (550-480 BC) Protagoras (485-410 BC) Brouwer (1881-1966) Piaget (1896-1980)
Realism (20the century) Scientific models are constructs designed to predict certain aspects of an objective reality that exists independently of the observer.	Emphasize the difference between models, which are constructed by scientists, and reality, which exists independently of models. Models are successive approximations of reality.	Aristotle (384-322 BC) Reid (1710-1796) Planck (1858-1947) Russel (1872-1970) Einstein (1879-1955)

Source: Riopel, Martin. 2005. Epistemology and science education. Université du Québec à Chicoutimi, 30p.

3. Comparative analysis of three doctoral theses

3.1. Summary presentation of each thesis

3.1.1. Law school thesis

The first thesis that caught our attention was that of Donald Poirier, published in 1993 by the Faculty of Law of the Université de Montréal. It is entitled: "*La protection juridique des personnes âgées. Étude socio-économique comparée au Nouveau-Brunswick et des provinces canadiennes*".

This thesis aims to explain the links between social actors and the adoption by the Atlantic provinces of legislation designed to protect the elderly from physical violence. The thesis is divided into three parts, each with at least three chapters. The first part presents the theoretical framework in the light of which the various legislative analyses and socio-legal methods were carried out. The second describes the evolution of the law relating to the protection of the elderly in Canada in the light of the theoretical framework developed by Michel Foucault, and the final part applies the theoretical framework to the legal protection of the elderly in New Brunswick. In keeping with the sociological approach, three aspects are studied: the creation of the legal norm, its interpretation by the courts and its application by social workers. Then comes the conclusion, which summarizes the study and poses a question to enable other researchers to explore other aspects related to the protection of the elderly, in particular their apprehension, as social actors, of the laws designed to protect them.

3.1.2. Faculty of Pharmacy thesis

The thesis entitled "*Matrices polymériques à gradient de concentration : formulation, évaluation et modélisation*" was written by Patrice Hildgen and published in 1995 at the Faculty of Pharmacy, Université de Montréal.

In this thesis, the author looks at the efficient pharmaceutical form to be adopted in drug manufacture. The author points out that the release of active ingredient from polymeric matrices is known to have Fickian kinetics, i.e. the amount of drug released depends on the square root of time. And the actual kinetics and simulations obtained by the algorithm enable us to better understand the close interrelationship between the dissolution and diffusion phenomena. As the author points out, of all pharmaceutical forms, the matrix form is the most advantageous and promising.

The book is divided into three parts. The first part deals with the theoretical framework. Here, the author traces the history and solutions relating to the realization of a concentration gradient. He focuses on the tools to be used, describes the experimental process and formulates his hypotheses. Attention is also paid to the variables and parameters identified in the system, in particular the independent and state variables, as well as the parameters of manufacture, constitution and in vitro experimentation. The author also describes his approach to demonstration. He opted for a macroscopic approach to solve the equation and a microscopic approach to characterize the porous media and analyze the image[1] .

In the second part, the author carries out experiments to verify his initial hypotheses. He constructs the automaton, presents the calculation algorithm, enters the parameters and proceeds with the tests.

Finally, in the last section, the author presents the results of the analysis and opens the discussion in the light of his initial hypotheses. He makes a number of suggestions, emphasizing the advantages of the matrix form.

3.1.3. Thesis of the planning faculty

The third thesis is by Mr. Marc-André Turcotte. Its title is "*Éléments d'analyse de la notion d'espace*". It was published by the Faculté de l'aménagement de l'Université de Montréal in 1980. In this thesis, the author sets out to demonstrate the close relationship that has historically prevailed between the space of symbolic representations and the space of rational representations.

This thesis also comprises three parts. The first part looks at the main conceptions of space, i.e. explicit conceptions, those of philosophy and physics. The second looks at implicit conceptions of space as applied to social space, the city, the urban, the regional and the national.

The final section is based on a number of analyses drawn from the history of mentalities, and seeks to illustrate the close relationship that prevailed at the origin of rational knowledge between representations of reality and the space of symbolic representations.

---

[1] Patrice Hildgen, *Matrices polymériques à gradient de concentration : formulation, évaluation et modélisation*, Montréal, Université de Montréal, 1995.

3.2. Problem analysis

3.2.1. Legal thesis

3.2.1.1. Clarity of the problem

In view of the path followed and the literary form adopted, the problematic of the thesis seems self-explanatory. The author starts from general considerations of personal protection, influenced by the civil law of European and Anglo-Saxon countries, to demonstrate the need to analyze the specific case of New Brunswick. He also focuses on the different age groups to demonstrate that people aged 65 and over constitute a category at risk.
In addition, the study examines the conditions that led to the identification of the elderly as a group of people at risk of abuse, and for whom protective legislative measures were consequently taken.

With regard to the different countries, the author refers to the very basis of the protection of incapacitated persons. On the one hand, he refers to the case of civil law countries (including Europe and Quebec), which prefer to maintain a law of general application, subject to the general principle that adults are presumed capable and incapacity must be demonstrated before they lose their right[2] . On the other hand, he points

[2] Donald Poirier, *La protection juridique des personnes âgées. Étude socio-économique comparée au Nouveau-Brunswick et des provinces canadiennes*, Université de Montréal, 1993.

to the case of Anglo-Saxon countries, which emphasize laws specifically designed to protect the elderly or people with disabilities. According to the author, these laws are based on a distrust of the family, seen as a major source of violence against individuals, and an extreme reliance on the social professions, deemed capable not only of diagnosing social problems, but also of remedying them in the interests of those entrusted to their protection (ref. New Brunswick). Unlike children, the elderly have legal capacity. Hence the need to protect them.

In addition, the author focuses on the description of the population. According to the author, in Canada, the under-16 population represents 19.5% of the total population, the over-65s 10.4%, and the mentally and physically handicapped 12%. In view of the growth in the various categories of the population, especially those aged 65 and over (which is expected to reach 10 to 20% in 2003), there is a need for protective legislation[3] .

However, this raises the issue of state interference in the lives of the elderly against their will. The author focuses on the question of legality and legitimacy. Indeed, the legitimacy of legislative activity and that of the courts, which attempt to reconcile the principle recognizing the autonomy of the human person and respect for the will of the legislator, with the legality that the state gives itself (the relationship between state and society), makes it possible to study the law relating to the protection of the elderly at each of its three stages, that of the development of the law, that of legal discourse and that of its implementation[4] .

---

[3] Donald Poirier, *La protection juridique des personnes âgées. Étude socio-juridique comparée du Nouveau-Brunswick et des provinces canadiennes*, Montréal, Université de Montréal, 1993, p. 6.
[4] Donald Poirier, *Op. cit.* p.7

3.2.1.2. Construction of the argument in relation to the theoretical framework

In terms of theoretical framework, the author refers to the sociology of law to show how older people are regulated. He endeavors to demonstrate that there is a relationship between legal and social rationalities.

Indeed, the sociology of law is a discipline whose principal aim is to develop our knowledge of the foundations of the legal order, the model of social change and the contribution made by law to the satisfaction of social needs and aspirations. According to the author, the followers of the sociology of law are opposed to those of legal positivism, who believe that only positive law exists, that law is a set of rules and that it forms a closed, complete and coherent system[5] .

Similarly, the sociology of law postulates a close relationship between society and law. Law is seen as one of the sub-systems of social regulation (all the material and symbolic resources available to a society to ensure that its members behave in accordance with a set of prescribed and sanctioned rules and principles). Thus, the social control exercised by law is fundamentally based on culture, on the existence of social norms and values that provide the link between living law and official law.

3.2.1.3. Construction of the argument in relation to the methodological framework

The author builds his argument on the approach developed by Michel Foucault, coupled with the methods of the human sciences, to flush out

---

[5] *Ibid.* p.18-24

the various logics at different moments in the law. He then analyzes the history of the protection of the elderly in Canada to highlight the characteristics of the legal logic of the State and that of the social logic conveyed by the social professions since the mid-nineteenth century, in order to understand how this social logic permeates the laws passed by the legislator[6] .

It also examines the interplay of social and legal norms at each of the three moments in the law relating to the protection of the elderly. Special emphasis is placed on New Brunswick legislation.

He also shows why he chose New Brunswick. According to the author, when the study aims to establish the conditions of emergence of a law and its implementation, it is appropriate to restrict oneself to a given territory over which a precise and unique political and legal sovereignty is exercised. As the Atlantic provinces are the only ones in Canada to have adopted legislation specifically aimed at protecting the elderly, he chose one of them (New Brunswick as the research target). The choice of this province is justified by the fact that the provisions relating to the protection of the elderly were adopted by the legislator in 1980, whereas the legislation of Nova Scotia (1985), Prince Edward Island in 1988 and Newfoundland in 1973. New Brunswick was also chosen because of its easy access to information and the researcher's familiarity with the region[7] .

Finally, the author discusses the usefulness of using qualitative methods in the implementation of the law. He indicates that field studies using

[6] Donald Poirier, *Op. cit.* p.40
[7] Donald Poirier, *Op. cit.* p.12

semi-structured interviews are used to verify how social workers put into practice the law relating to the protection of the elderly in New Brunswick.

3.2.1.4. Epistemological stance adopted

Rationalism

A close look at the problematic of this thesis reveals that the author is closer to epistemological currents such as rationalism, structuralism and positivism. According to *Le dictionnaire actuel de l'éducation*[8] , rationalism is an epistemological current that considers that "*all valid knowledge derives either exclusively or essentially from the use of reason*".

Indeed, the author refers to the theoretical model based on the notion of rationality to explain the transition from social practice to legal rationality. According to Ewalt, quoted by the author, through the notion of rationality, it becomes possible to resolve the old problem of being and ought-to-be. The concept of rationality is thus a means of grasping the articulation between different legal systems, and between law and other types of social practice[9] .

Through the study of rationalities, it becomes possible to discover the various social practices at work in the different moments of the law, whether in its emergence, in its implementation or in the law as a legal

---

[8] *Le dictionnaire actuel de l'éducation*, 1994, p.1003
[9] Donald Poirier, *Op. cit.* p. 53

norm to be interpreted, and thus to understand how they influence each other.

Structuralism

The author builds his theoretical framework mainly on the ideas put forward by the structuralist school, notably Michel Foucault. Indeed, Michel Foucault developed both a historical and epistemological approach to understanding contemporary social phenomena. This approach aims to uncover the new types of rationality that have emerged and their multiple effects[10] . He is particularly interested in the evolution of social regulation and the points where the different stages of this evolution meet. He found that *power and knowledge* are inseparable. Power makes it possible to obtain certain types of knowledge. This knowledge then generates its own objects of knowledge[11] .

The Foucauldian approach to law is twofold: *genealogical*, i.e. tracing the genealogy of the law back to its origins; and *archaeological*, i.e. understanding how such a problematization of the issue of protecting the elderly came about[12] .

With regard to the theory of social control, the author cites Black and Foucault. According to the latter, social regulation by the state increases in proportion to the weakness of the private system of social regulation constituted by the family, the church, the school and other organizations. In other words, when informal mechanisms of social regulation are

[10] Donald Poirier, *Op. cit.* p. 60
[11] *Free encyclopedia Wikipedia,* http://fr.wikipedia.org/wiki/Michel_Foucault
[12] *Ibid.*

strong, there is no need for a highly developed legal system of social regulation[13].

Finally, the reference to structuralist methods in the study of law is an attempt to understand how judges reconcile the rationality of the social sciences with traditional legal rationality.

Positivism

The positivist movement is inspired by empiricism in that it focuses on observational facts alone, but recognizes the importance of reasoning, adding that the sciences use mathematization to link experimental data together as simply as possible. It should be noted that positivists insist on the rigor of inductive reasoning, which enables us to move from facts to hypotheses[14].

The study borrows most of these concepts from the paradigm of *legal pluralism.* This paradigm postulates the coexistence of a plurality of legal frameworks or systems within a given unit of sociological analysis. The author endeavours to demonstrate that the development and application of law are essentially dynamic social processes, marked by an incessant dialectical interaction between the formal and the informal, on the one hand, and between the various private and public components of a society's own system of social control, on the other[15],

3.2.1.5. Contributing to the advancement of knowledge

---

[13] Donald Poirier, *Op. cit.* p. 40-46

[14] *Martin Riopel, Epistemology and science education, http://www.astro.umontreal.ca/~riopel/epistemologie/*

[15] Donald Poirier, *Op. cit.* p. 24

The study sheds some light on the nature of the law. No research on the courts' interpretation of elder protection legislation existed in Canada. Even in the U.S., according to the author, little research has been done on the judicial interpretation of elder protection laws, and very little research has sought to demonstrate the mechanism by which social logic has found its way into the legal system.

In addition, the research points to the fact that the social logic introduced into laws designed to protect the elderly is inadequate to solve the problems targeted by these laws, and inadequate to solve the sheer number of legal problems faced by the elderly. Contractual solutions based on legal logic are examined.

3.2.2. Pharmacy thesis

3.2.2.1. Clarity of the problem

Examination of the problematic shows a coherent description of the problems posed by drug release. The literary form adopted also makes the problem explicit.

Indeed, the author points out that, despite having the parameters, the major problem in drug manufacture is the inability to predict the exact kinetics of matrix formation. With a view to indicating what has been done and what remains to be done, he presents the work carried out on this issue since the 1960s. It also lists the authors who have contributed to the elucidation of the physico-chemical phenomena governing drug release into the surrounding environment.

In addition, he uses certain theories that have already been validated to demonstrate the interest of his research. Such is the case of Broadbents and Hammersley's theory of invasion of porous or potentially porous media. The same is true of T. Higuchi's work on mathematical modeling of the release kinetics of pharmaceutical forms[16] .

3.2.2.2. Construction of the argument in relation to the theoretical framework

To build his theoretical framework, the author refers to the history and solutions already proposed. In particular, he discusses the physical laws, mathematical principles and existing models that have tried to predict and propose solutions to the problem posed.

Indeed, the release of the active ingredient is physically described by more than one phenomenon, all of which are governed by physical laws that cannot all be expressed simply[17] . With regard to drug release, in the case of a polymeric matrix for example, he identifies two physical phenomena: dissolution and diffusion. With regard to solutions already proposed, the author refers to the modeling of active ingredient release in forms with a concentration gradient, as postulated by Chang and Himmelstein, who suggested as a solution a numerical integration method, assuming that diffusivity is a simple function of time. He also points to Higuchi's work on mathematical modeling of the release kinetics of pharmaceutical forms[18] .

[16] Patrice Hildgen, Matrice polymériques à gradient de concentration : formulation, évaluation et modélisation, Montréal, Université de Montréal, 1995, p.10

[17] Patrice Hildgen, *Op. cit.* p. 31

[18] *Ibid.* p.15

As for the problem posed by dissolution (the invasion of the porous medium by the medium, i.e., a hydrodynamic phenomenon), the author demonstrates the need to resort to Fick's model, as he points out that the latter assumption cannot easily be verified experimentally.

3.2.2.3. Construction of the argument in relation to the methodological framework

To build his argument in relation to the methodological framework, the author defines the necessary variables and parameters to be brought into play during experimentation.
Indeed, before starting any modeling, it's a good idea to identify and define all the system variables and parameters you're likely to encounter. It defines the independent variables, the state variables and the main parameters, i.e. manufacturing parameters (temperature, centrifugation speed, polymer viscosity, centrifugation time, etc.), build-up parameters (matrix length, total amount of principle, etc.), and the process parameters (temperature, centrifugation speed, polymer viscosity, centrifugation time, etc.).These include manufacturing parameters (temperature, centrifugation speed, polymer viscosity, centrifugation time, etc.), constitutive parameters (matrix length, total quantity of active ingredient in the matrix Q, K the dissolution constant, Ø the solubility of the drug in the medium, fip the solubility of the drug in the polymer) and in vitro experimental parameters (temperature during release, also kept constant as its influence on the diffusion coefficient and solubility is very great, PH of the release medium). These main parameters enable him to develop his equation.

He then indicates the experimental procedures and approach to be used to verify his initial hypotheses. In this respect, he discusses the macroscopic approach to solving the equation, and the microscopic approach in the context of physics close to statistical physics. Given the complexity of the diffusion path, the author takes care to incorporate a correction factor into his model, namely Turtuosity. This correction factor was defined by Higuchi and is directly related to the geometry and topology of the medium.

Finally, the author outlines the qualities of the instruments he intends to use in the experiment and minimizes any shortcomings. He points out, for example, that the best tool for describing complex geometry is fractal geometry, since the dimensions it can define are effective for describing a porous matrix[19] .

3.2.2.4. Epistemological stance adopted

We can situate the author within the *positivist* movement . The positivist movement is inspired by empiricism in that it focuses on observational facts alone, but recognizes the importance of reasoning, adding that the sciences use mathematization to link experimental data together as simply as possible. It should be noted that positivists insist on the rigor of inductive reasoning, which enables us to move from facts to hypotheses. As inductive reasoning is indispensable (for positivists) to the evolution of science (according to Auguste Comte's famous formula "*seeing is*

[19] Patrice Hildgen, *Op. cit.* p.57

predicting"), the theories produced have no value in themselves other than that of being linked to facts[20] .

In fact, he constructs his problem based on various laws and theorems relating to mass and speed. These include Newton's law, Fick's law, Falconer's theorem, etc.[21] . He also drew on a number of previously validated theories, notably the percolation theory defined by Broadbents and Hammersley in 1956 and formalized by Gennes in 1976 and Staffer in 1985, which describes the transport properties of matter in heterogeneous media[22] . He then proceeds to experiment, discusses his results and formulates a general proposal to describe the fastest process for drug release in the context under consideration.

3.2.2.5. Contributing to the advancement of knowledge

The author shows a real desire to contribute to the advancement of knowledge by proposing a different approach to drug production, based on physics close to statistical physics (microscopic approach).

In addition, to address the problem of increasing path length in concentration gradient polymers, his thesis provides a method that saves time by predicting in vitro kinetics, and allows the optimal formulation to be reached quickly.

---

[20] Martin Riopel, Epistemology and science education, http://www.astro.umontreal.ca/~riopel/epistemologie/index1.htm
[21] Patrice Hildgen, *Op. cit.* p. 34-38
[22] *Ibid.*

3.2.3. Planning thesis

3.2.3.1. Clarity of the problem

A careful reading of the thesis provides information on the coherence of the thesis development in general and the construction of the problematic in particular. The elements below lead us to maintain that the thesis' problematic is explicit.

Indeed, the model of interpretation developed by the author is based on a number of concepts: the social imaginary, the historical, the symbolic and the rational. The author refers to the historical relationships between the realities summarized by the concepts of space (real space, symbolic space, social space) to demonstrate the interpretation given to the notion of space at different moments in history. He starts from the very foundation of the conceptualization of man's particular relationship with his environment, to describe the purely representational character of the various conceptions of space. To this end, he poses the problem of perception and interpretation of the real: real, cosmic, cultural and mental space.

With regard to real space, for example, he points out that these representations are based on models of analysis most often inspired by previous knowledge, i.e. philosophy and/or physics, and these conceptions are generally expressed in an implicit mode[23] .

---

[23] Marc-André Turcotte, *Éléments d'analyse de la notion d'espace*, Montréal, Université de Montréal, 1980, p.4-30

On the subject of social space, the author alludes to the technotopic and anthropocultural versions of the various representations of social space. In the first case, social space is reduced to the behavior of a more or less inert mass, or to the dynamics of energy (matter more or less subject to human action), an interpretation that aims to neutralize an essentially cultural phenomenon. In the second case, the naturalization of cultural space is elaborated both more openly and more confusedly, on the basis of a reduction of culture to human nature[24] .

According to the author, these two modes of naturalization - sometimes by reducing the city to a more or less inert mass, sometimes by establishing some kind of functionality of the human body, of its psyche with social space - aim to give the latter a legitimacy, that of a new ecological order, essentially anthropomorphic, a kind of second-degree natural data[25] .

3.2.3.2. Construction of the argument in relation to the theoretical framework

With regard to the theoretical framework, the author constructs his problematic on the basis of the main conceptions of space, from those promoted by philosophy to those developed by classical and modern physics. What these conceptions of space have in common, he says, is that they expressly or explicitly propose a definition of space that is above all coherent or rational. Both philosophical and physical conceptions rely on rational demonstration to acquire the status of truth, of exact and authentic knowledge of reality in general.

---

[24] Marc-André Turcotte, *Op. cit.* p. 12-20
[25] *Ibid.*

Indeed, the philosophical approach confronts two schools of thought. The first sees space as an absolute entity, independent of the cosmic matter it may contain, while the other view sees space as an entity relative to the elements it contains[26] .

Speaking of Einstein's conception of space, for example, the author points out that his conception is akin to absolutist conceptions of space insofar as a formal framework, independent of matter energy, is detected in the elaboration of the General Theory of Relativity (GRT)[27] .

Finally, the author draws on anthropological considerations. He stresses that physics, like philosophy, far from leading to immutable data about space, reveals a historically determined anthropological content[28] .

3.2.3.3. Construction of the argument in relation to the methodological framework

With regard to the methodological framework, the author deals with the notion of space by considering its historical, rational and social aspects. He takes an inductive approach to explaining the notion of space. Through history, he endeavors to demonstrate the interpretation given to the notion of space (real space, symbolic space, relative space, mathematical, linguistic and communicational space, social space).

Indeed, space is conceived as an entity radically different from the things that make up reality (Newton). Leibniz advocates the prima of absolute

[26] M arc-André Turcotte, *Op. cit.*, p.33
[27] *Ibid*, p.34
[28] Ibid.

space. Physics therefore acknowledges compromises with a certain imaginary and formalizing order. Space is therefore first and foremost an anthropological datum, and not strictly objective or self-existent[29] .

Moreover, in the process of language genesis, he cites the two-axis theory developed by Kraszewski, Saussure and Jakobson, highlighted by linguistics and semiology.

3.2.3.4. Epistemological stance adopted

The author draws on several epistemological currents to elucidate the notion of space (rationalism, historicism, positivism, idealism and materialism, constructivism, realism, humanism).

Rationalism and Historicism

Rationalism is an epistemological current which holds that "all valid knowledge comes either exclusively or essentially from the use of reason. *Historicism* is based on the idea that truth is the result of a self-engendering of reason, whose historical development is revealed in *moments*, all necessary but insufficient[30] .

Historically, the knowledge associated with the field of geometry has played an important role in the development and justification of the rationalist epistemological position. The inventor of analytic geometry, the French mathematician René Descartes, quoted by the author, took up

[29] Marc-André Turcotte, *Op. cit.* p.74
[30] Free encyclopedia Wikipedia, *http://fr.wikipedia.org/wiki/Historicisme*

the rationalist position by attempting to apply the rigor and clarity of mathematics to the field of philosophy.

Idealism and materialism

Absolute space

Materialism refers to a set of philosophical doctrines which, rejecting the existence of a spiritual principle, reduces all reality to matter and its modifications. The first, the theory of absolute space, maintains that among the elements of the (ontological) reality of the universe, there exists an entity called space. According to this approach, space cannot be seen as or reduced to a set of relations, of material things, but is in fact ontologically prior to the class of all material objects (sensible and supra-sensible reality). Space is then conceived as a substance. Space is conceived as an entity radically different from the things that make up reality (Newton).

Relative space

Idealism, on the other hand, is characterized by the affirmation of an intelligible reality (realism of the intelligible), the assertion that thought is the only certain reality, the reduction of being to thought or consciousness, the reduction of phenomena to representations of the mind, and the derivation of beings and reality as a whole from a spiritual principle (thought, consciousness, concept, etc.)[31] . In his work, the second theoretical approach, relativism, considers space as a set of relations between material things. This second approach excludes any

[31] Free encyclopedia Wikipedia, http://fr.wikipedia.org/wiki/Id%C3%A9alisme

possibility of space being a subsistence, an ontological entity. Both conceptions are contested[32] . Leibniz even speaks of the primo of absolute space[33] .

Positivism

This trend suggests that the mind gives up the "why? question, i.e. the search for an absolute explanation of things. It would limit itself to the "how", i.e. to formulating the laws of nature, by identifying, through repeated observations and experiments, the constant relationships that unite phenomena[34] .

According to the author, Earman indicates that there is a Newtonian absolute space quite similar to Leibnizian relative space, and the existence of space, however conceived, derives more from the mode of intervention of the analyzing subject and the context that gives rise to this type of intervention than from the specificity of the object analyzed (notably the materiality of space)[35] .
At the same time, physics recognizes compromises with a certain imaginative and formalizing order. Space is therefore first and foremost an anthropological datum, and not strictly objective or self-existent[36] .

The author also alludes to Einstein. Einstein's geometric description of physical theory has its origins in the advances of non-Euclidean geometry, which can be traced back to the various attempts over the

---

[32] Marc-André Turcotte, *Op. cit.* p.37
[33] *Ibid.* p.54
[34] Free encyclopedia Wikipedia, http://fr.wikipedia.org/wiki/Positivisme
[35] Marc-André Turcotte, *Op. cit.* p.55
[36] *Ibid.* p. 74

centuries to prove Euclid's fifth postulate, stating that a point can only lead parallel to a given line.

Whereas the philosophical conception engages a purely voluntarist and idealistic approach, i.e. a desire to transfer the whole of the cosmos within the limits of a discourse, physics, for its part and in contrast to the former, finding both refuge and inspiration in the austere universe of mathematics, claims first to deduce its model within the framework of a praxis that engages it directly in the work of systematic observation of nature.

Constructivism

Constructivist epistemology poses the problem of objectivity. In the first case, the active role of the subject (the philosopher) often confuses the position of the object (Reality). In the second case, the epistemological break is the very principle of the operation, with the mathematical instrument playing a central role by (apparently) consecrating the neutrality of the observer. There is a danger of formalization[37] . For Jean Piaget (1896-1980), "*a fact is always the product of composition, between a part provided by the objects, and another constructed by the subject*"[38] . In this thesis, the elements below lead us to orient the author towards the constructivist current. According to this current, from Freudian analysis to Jean Piaget's genetic epistemology, via Bourbaki's mother structures, no convincing thesis has been produced as to the intra-organic, psychological or ontological foundation of the elementary categories of knowledge. In contrast to this trend, others have sought to establish this

[37] *Ibid.* p. 78
[38] Jean Piaget, *Logique et connaissances scientifiques,* p.

foundation within the framework of a reflexive relationship between the mind, the brain or consciousness and reality, and more specifically the social[39] .

Realism

Realism recognizes the existence and primacy of a reality that is independent of any observer, objective and absolute, immanent. This is why realism tends to affirm the existence of a reality in itself, structured according to intelligible archetypes such as ideas [40]. Most currents of thought ultimately recognize a certain relationship between reality and subjectivity[41] . In his analysis of the notion of space, the author identifies the following paradoxes:

The first paradox concerns digital communication and its relationship to the analog. The distinction between analogue and digital depends very much on how we define the distinction. As systems theory makes clear, we cannot lose sight of the fact that we are talking about models constructed for the purposes of explanation, not natural models, since nature does not seek to explain[42] . The second is linked to Goltlob Frege's concept of zero. According to Frege, zero is an objective concept. But the object is not necessarily linked to reality, to existence, to space, because each object does not necessarily have a place (or at least, a geometric link)[43] . The third relates to the set Ǿ and the Euler diagram. Here, the author refers to set theory, which states that the complement of any subset A in a set S is the subset B. The empty set is a unique set that contains no

[39] Marc-André Turcotte, *Op. cit*, p.102
[40] http://fr.wikipedia.org/wiki/R%C3%A9alisme#Principes_du_r.C3.A9alisme
[41] Marc-André Turcotte, *Op. cit*, p.103
[42] *Ibid.* p. 91
[43] Patrice Hildgen, *Op. cit.* p. 92

elements. Finally, the fourth paradox deals with Russell and Whitehead's logical types

Humanism

Humanism refers to all thought that places the development of man's essential qualities at the forefront of its concerns, and denounces that which enslaves or degrades him[44] . Making man the point of convergence of the Real, the apparently conflicting discourses of idealism and materialism ultimately espouse the same metaphysical project, that of making Man the point of fall for emancipation from the animal condition[45] . According to the author, this undertaking to transform the Real, which is favorably endorsed by this humanism, takes shape in the social space, which constitutes the most effective material support. Speaking of the idealist critique, the author stresses that this remains without doubt Marx's most decisive work[46] .

Urban space, in particular, takes on the value of a dynamic, objective symbol of this enterprise of domination, of the subjugation of natural forces for the main benefit of humankind, for which idealist or materialist discourse offers an absolute explanation. These fragile philosophical foundations form the basis for two main types of analysis: the technotopia and the anthropocultural approach.

However, neither is explicitly based on any conceptualization of the Real in terms of space. Both have the social space, the city, as their sole field of vision, and are most often based on an implicit conception of space.

---

44 http://fr.wikipedia.org/wiki/Humanisme
45 Patrice Hildgen, *Op. cit.* p.104
46 *Ibid*, p.107

According to the author, the disciplines involved in the production of urban space - architecture, urban planning and, more generally, regional planning - are as concerned with the form that socio-cultural space takes as they are with the "conscious/unconscious" division of the Real, which in certain respects is reminiscent of the spell cast by absolute philosophers and physicists[47] .

3.2.3.5. Contributing to the advancement of knowledge

The book contributes significantly to an epistemological reflection on the use of the notion of space, and provides a synthesis of the main conceptions of space. The thesis also sheds light on the problem of the historicist reduction of urban social space to an evolutionary and ontological phenomenon[48] .

3.3. Main research question

3.3.1. Legal thesis

The study attempts to explain the links between social actors and the adoption by the Atlantic provinces of legislation designed to protect the elderly from physical violence. Several questions arise: - What conditions favoured the emergence of laws to protect the elderly? - What values does the legal discourse convey? How are these laws implemented, and what are the consequences? - Under what conditions is the state authorized to limit the freedom of movement of the elderly? - What rationales lie behind laws designed to protect the elderly? - How have non-legal

---

[47] Patrice Hildgen, *Op. cit.* p. 104-105
[48] *Ibid.* p. 31

rationales found their way into legal discourse? - Are there conflicts between legal rationality, which recognizes the principle of personal autonomy, and social rationality, which insists on protection against the risk of violence to which the elderly are subject? - What effect do these laws have on the elderly as a group, on the people who look after them, on the rights and freedoms of citizens and, finally, on our conception of violence (socio-legal approach)?

The main question seems representative of the epistemological posture, as it seeks to establish the links between legal rationality and social rationality. It fits coherently into the problematic as a whole, as can be seen from the various sub-questions the author poses.

3.3.2. Pharmacy thesis

The main research question is how to remedy the problem of the increased distance to be covered in gradient polymers in the drug manufacturing process, taking into account the different pharmaceutical forms?

According to the author, despite having the parameters, the major problem in drug manufacturing is the inability to predict the exact kinetics of matrix formation. He distinguishes two aspects to this problem: a technological aspect, to design a system that meets the requirements, and a theoretical aspect, to have a mathematical model that enables the system's operation to be predicted. This question is representative of the epistemological stance adopted.

3.3.3. Planning thesis

The main research question is how to identify, with the help of new observations derived from a reconsidered historiographical approach, a few key facts that establish a relative continuity between different conceptions of space? The study therefore seeks to identify the close relationship that has historically prevailed between the space of symbolic representations and that of rational representations, and to bring out through analysis the main representations of space and the determining aspects of the different representations of the notion of space. The research question seems representative of the epistemological posture.

3.4. Comments

3.4.1. Similarities

For all three theses, each researcher sets out to define the parameters of his or her research. Emphasis is placed on the object to be investigated, which the project sets out to circumscribe by setting an investigative objective, linked as appropriate to description, classification, experimentation or modelling.

The researcher also defines the empirical and conceptual framework likely to give the research its full relevance, through reference to similar studies already carried out, as well as to the concrete issues underlying the problem being studied.

The problematic, the point of view from which the object is apprehended, a suspensive and interrogative point of view, itself specified by a body of hypotheses, depending on the situation, is philosophical or theoretical in nature, or disciplinary or empirical. Similarly, each author adopts a specific approach to achieve his or her research objective. The methodology (approach, methods and techniques) specifying the conduct of the research process and the planned means of investigation is highlighted.

Finally, each author's research project is an attempt to move imperceptibly from the given object, a real object defined thematically, to the constructed object, an object freed from prejudice, no longer expressible thematically but as an operative concept that can be directly put to the test of facts[49] . Each thesis strives to add "a brick" to scientific knowledge. However, for scientific knowledge to exist, there must be a project that constructs an object where theory and experiment are linked in a verification[50] .

3.4.2. Differences

We'll refer to Jean-Pierre Boutinet's book on project anthropology to identify the differences between a thesis in must, pharmacy or planning.
In his legal thesis, the author examines the conditions for the emergence of protection for the elderly in New Brunswick. According to Jean-Pierre Boutinet, three elements characterize a bill:

- the need to explain the reasons for taking action, taking into account the previous situation; a draft law not accompanied by reasons will be deemed unconstitutional
- ongoing collective bargaining; and

[49] Jean-Pierre Boutinet, *Anthropologie du projet*, Paris, PUF, 2003, p.106

[50] *Ibid*. p.107

- the fact that the author of the bill is also, in one way or another, its producer (unlike in the case of the proposal)[51] .

In addition to adopting a deductive approach, the legal thesis is inspired by rationalist, structuralist and positivist epistemological currents. Indeed, the author refers to a theoretical model based on the notion of rationality to explain the transition from social practice to legal rationality. Max Weber developed the notion of rationality. According to Ewalt, through the notion of rationality, it becomes possible to resolve the old problem of being and ought-to-be. The concept of rationality is thus a means of grasping the articulation between different legal systems, and between law and other types of social practice.

It also draws on Michel Foucault's method. Foucault developed both a historical and epistemological approach to understanding contemporary social phenomena. The study borrows most of its concepts from the paradigm of *legal pluralism.* This paradigm postulates the coexistence of a plurality of legal frameworks or systems within a given unit of sociological analysis.
Similarly, the author attempts to demonstrate that the development and application of law are essentially dynamic social processes, marked by an incessant dialectical interaction between the formal and the informal, on the one hand, and between the various private and public components of a society's own system of social control, on the other.

It's worth noting that the law dissertation from a law school responds directly to a need, which is the protection of the elderly. It draws on more than one discipline, notably law, sociology, psychology and economics.

[51] Jean-Perre Boutinet, *Op. cit*, p.111

For the pharmacy thesis, the approach seems less complex. From the observation of specific facts, the author infers a general proposition or law. There are certain fixed steps that the researcher must follow in the experimental process. Testing is essential to validate the study. These steps include: defining the problem, formulating the hypotheses, describing the experimental procedure, identifying the variables and parameters to be used in the experiment (manufacturing parameters, constitution, experimentation), formulating the equation, introducing a correction factor, defining the qualities of the experimental instruments (condition, advantages), constructing the model (equation), defining the correction parameters, presenting the results and stating a law.

According to Jean-Pierre Boutinet, experimental projects are designed to mobilize energies around a challenge. Experimental organizations make innovation the decisive parameter. A need to enhance human and technical capital to maintain a sufficient level of inventiveness[52] .
From an epistemological point of view, the author is close to the positivist movement. His approach is inductive.

As for the planning thesis, its specificity lies in its interdisciplinary nature and the analysis of the history of different interpretations of the notion of space.

At , talking about real space for example, the author points out that the various representations of space are based on models of analysis most

[52] Jean-Pierre Boutinet, *Op. cit*, p.117

often inspired by previous knowledge, either philosophy and/or physics, and these conceptions are generally expressed in an implicit mode.

On the subject of social space, he evokes several conceptions of social space that have been initiated by sociology, economics, psychology, the sciences of spatial planning, anthropology and so on.
For example, the contribution of economic analysis can be upstream of a public or private decision-making process, and downstream of the analysis of an environmental problem by the physical and natural sciences. To be relevant, planners need to integrate elements from a wide range of disciplines, from law and engineering to ecology and hydrology.

What's more, most environmental problems are characterized by high levels of uncertainty, long-term effects, and a distribution of their impacts in space and time that is rarely uniform. It is on these essential points that interdisciplinary dialogue can be most fruitful, to best appreciate the margin of uncertainty affecting the expected effects of a given policy. Depending on the situation, dialogue with other disciplines can range from a simple exchange of information to the design of integrated models. Hence the use of several epistemological currents, notably rationalism, historicism, positivism, constructivism and so on. In the planning thesis, an effort is made to explain its theoretical framework.

However, it should be stressed that this planning thesis seems to us to be more philosophical. It is essentially concerned with the theoretical analysis of the notion of space. We agree with Jean-Pierre Boutinet's assertion that planning must aim for collective appropriation of geographical space through the mediation of technical achievements: construction of technical devices, modification of spatial configurations, all spatial initiatives aimed at facilitating this appropriation[53] . Planning implies intervention in the environment. It's a never-ending activity, to be repeated over and over again to create a more habitable space. At the very least, this activity implies taking into account three central parameters: ongoing negotiation between the various bodies of the community seeking to control its space; time, with its deadlines and indeterminate horizon, which disqualifies everything that is of the order of the punctual and the immediate; and space, with the identification of constraints, possibilities and everything that constitutes its singularity[54] .

The study of questions of fact, which includes the study of interactions between subject and object, never succeeds in confining itself to the field of facts alone, since everything that concerns the subject takes on a normative aspect in its consciousness, and therefore also falls within the scope of the preceding considerations.

Even in the field of purely experimental knowledge, questions of fact and validity are perpetually intertwined, requiring similar coordination of methods of formalization and historical and genetic reconstruction.

The close interdependence of direct, formalizing, historical-critical and genetic analyses stems from the fundamental need for a dialectic of genesis and structure, corresponding to their effective and alternative interactions.

Ultimately, we need to emphasize the complementary roles of inductive reasoning, deductive reasoning and experimentation in the search for new

[53] Jean-Pierre Boutinet, *Op. cit.* p.105
[54] Jean-Pierre Boutinet, *Op. cit.* p.104

scientific knowledge. There is a difference between models (which are produced by scientists) and reality (which exists independently of models), and we need to recognize a subjective and creative component in the development of scientific theories.

Conclusion

The study of questions of fact, which includes the study of interactions between subject and object, never succeeds in confining itself to the field of facts alone, since everything that concerns the subject takes on a normative aspect in its consciousness, and therefore also falls within the scope of the preceding considerations.

Even in the field of purely experimental knowledge, questions of fact and validity are perpetually intertwined, requiring similar coordination of methods of formalization and historical and genetic reconstruction.

It is useful to emphasize the complementary roles of inductive reasoning, deductive reasoning and experimentation in the search for new scientific knowledge. There is a difference between models (which are produced by scientists) and reality (which exists independently of models), and we need to recognize a subjective and creative component in the development of scientific theories.

The specificity of the planning thesis lies in its interdisciplinary nature and the analysis of the history of different interpretations of the notion of space.

Most environmental problems are characterized by high levels of uncertainty, long-term effects, and a distribution of impacts in space and time that is rarely uniform. It is on these essential points that interdisciplinary dialogue can be most fruitful, to best appreciate the margin of uncertainty affecting the expected effects of a given policy.

Depending on the case, dialogue with other disciplines can range from a simple exchange of information to the design of integrated models. Hence the need to draw on several epistemological currents, including rationalism, historicism, positivism, constructivism, etc., and to describe one's theoretical framework clearly.

1. Barreau, H., *L'épistémologie*, Paris, PUF, 127 p., 1995.
2. Begin, R., *Conception de la science et intervention pédagogique*, Spectre, vol.26, n°2, p. 10-16, 1997.
3. Boutinet, Jean-Pierre, *Anthropologie du projet*. Paris, PUF, 2003. 350p.
4. Dolle, J-M., *Pour comprendre Jean Piaget*. Paris, Éditions Dunod, 1989. 275p.
5. Hildgen, Patrice, *Concentration gradient polymer matrices: formulation, evaluation and modeling*. Montréal, Université de Montréal, 1995.
6. Galilée, G., *L'Essayeur*, Paris, Les Belles Lettres, 307 p., 1990.
7. Giordan, A. and De Vecchi, V. *Les origines du savoir*, Paris, Delachaux et Niestlé, 212 p., 1987.
8. Granger, G.-G., *La science et les sciences*, Paris, Presses Universitaires de France, 127 p., 1995.
9. Jarroson, B., *Invitation à la philosophie des sciences*, Paris, Seuil, 234 p., 1992.
10. Kremer-Marietti, A., *Le positivisme*, Paris, PUF, 128 p., 1993.
11. Largeault, J., *L'intuitionnisme*, Paris, PUF, 135 p., 1992.
12. Larochelle, M. and Desautels, J., *Autour de l'idée de science*, Sainte-Foy, Presses de l'Université Laval, 314 p., 1992.
13. Le Moigne, J.-L., *Les épistémologies constructivistes*, Paris, PUF, 127 p., 1995.
14. Newton, I., *The Principia: Mathematical Principles of Natural Philosophy*, University of California Press, 974 p., 1999.

15. Piaget, Jean & al, Logique et connaissance scientifique, Encyclopédie de la pléiade. Éditions Gallimard, Paris, 1967. 1345p.
16. Poirier, Donald, *La protection juridique des personnes âgées. Étude socio-juridique comparée du Nouveau-Brunswick et des provinces canadiennes.* Montréal, Université de Montréal, 1993.
17. Popper, K., *The Logic of Scientific Discovery*, Paris, Payot, 480 p., 1973.
18. Riopel, Martin, *Épistémologie et enseignement des sciences,* 2005. Université du Québec à Chicoutimi, 30p. http://fr.wikipedia.org/wiki/%C3%89pist%C3%A9mologie#Th.C3.A8mes_.C3.A9pist.C3.A9mologiques, accessed December 24, 2023
19. Robert, P., *Le nouveau petit Robert*, Paris, S.N.L., 2551 p., 1995.
20. Turcotte, Marc-André, *Éléments d'analyse de la notion d'espace.* Montreal: Université de Montréal, 1980.

MIX
Papier aus verantwortungsvollen Quellen
Paper from responsible sources
FSC® C105338

Printed by Books on Demand GmbH, Norderstedt / Germany